THE CENTRE OF THE EARTH

Leon

Hetty

Professor Verne

Brigit

Toosant

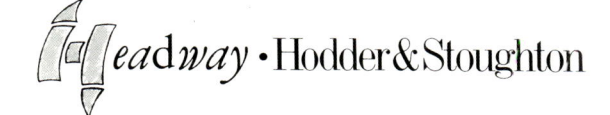

Newton

Headway · Hodder & Stoughton

In the nineteenth century Jules Verne, the famous author, wrote exciting books about journeys of adventure. Now his namesake, Professor Bartholemew Verne, has invented an amazing machine. In it he can follow the journeys Jules Verne wrote about. With his crew of children, he can explore the wonders of the world and of outer space.

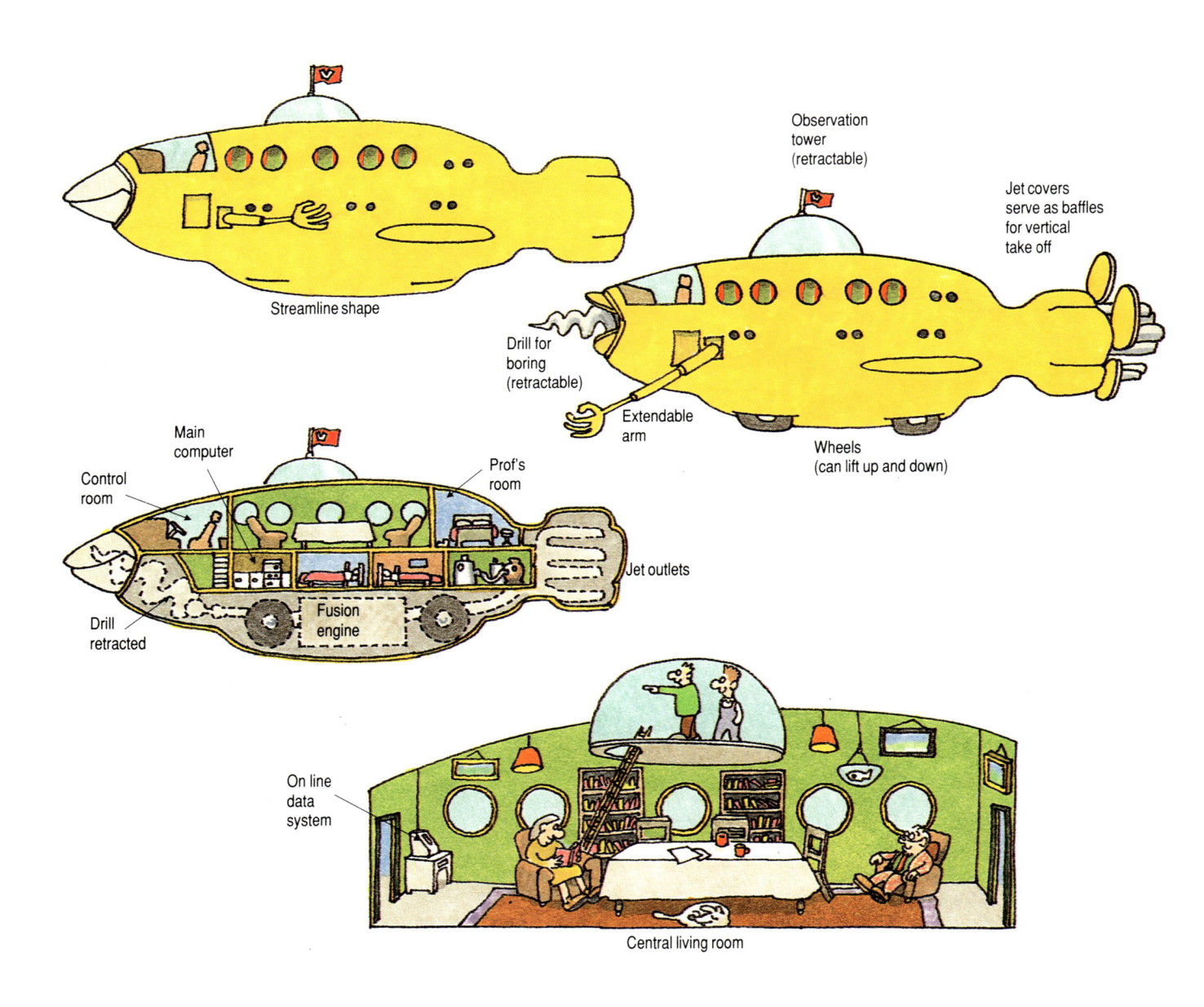

Streamline shape

Observation tower (retractable)

Jet covers serve as baffles for vertical take off

Drill for boring (retractable)

Extendable arm

Wheels (can lift up and down)

Main computer

Control room

Prof's room

Jet outlets

Drill retracted

Fusion engine

On line data system

Central living room

Right, crew. A hundred years ago Jules Verne wrote a book called 'JOURNEY TO THE CENTRE OF THE EARTH'.

On this journey we're going to find out what the inside of planet Earth is really like.

You mean — underground?

Yes, we're going to travel to the centre of the Earth!

!

Our vehicle can withstand the highest temperatures and the greatest pressures.

Thank goodness!

We're going to need all the protection we can get. But first, let's take a look at where we're going...

Destination: # PLANET EARTH

Mass: 5.975×10^{27} grams
Age: 4,600 million years

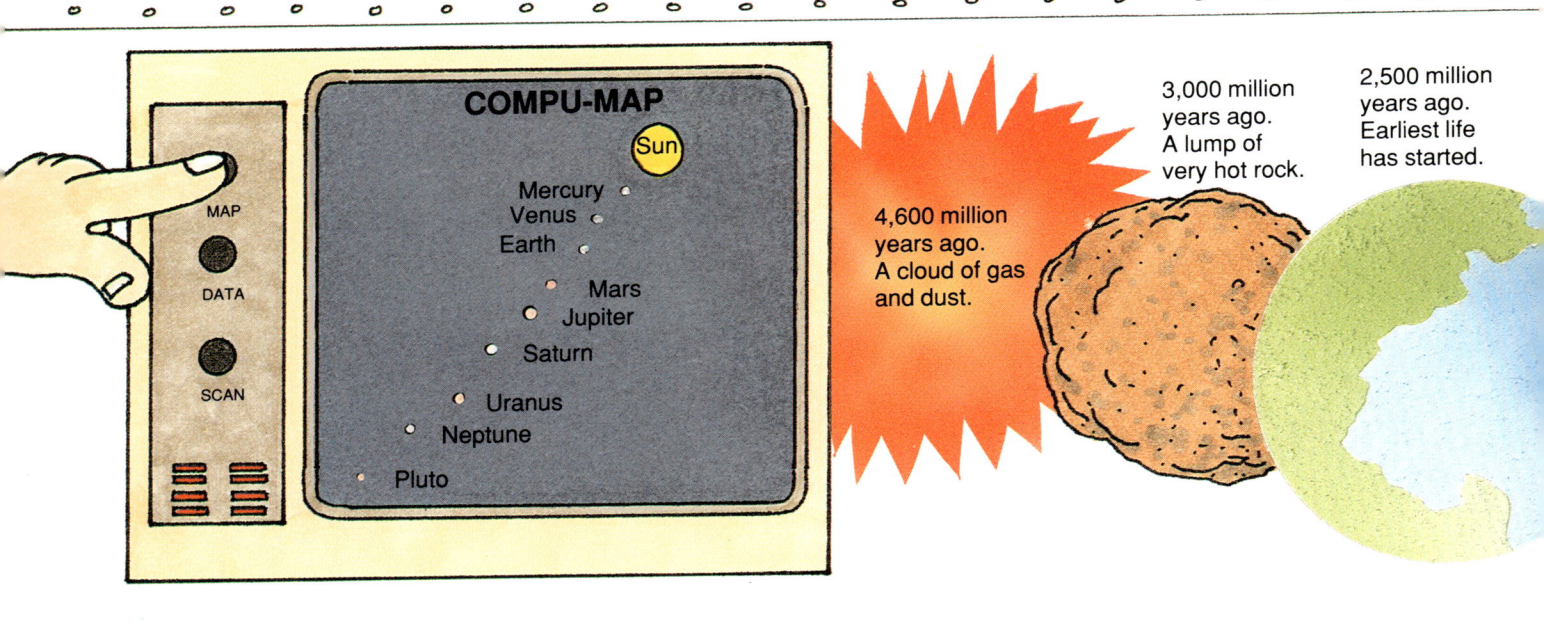

COMPU-MAP

Sun
Mercury
Venus
Earth
Mars
Jupiter
Saturn
Uranus
Neptune
Pluto

MAP
DATA
SCAN

4,600 million years ago. A cloud of gas and dust.

3,000 million years ago. A lump of very hot rock.

2,500 million years ago. Earliest life has started.

Earth is the third planet out from the sun. Scientists think that the sun and the planets formed at the same time from a huge spinning cloud of gas and dust.

The gases and dust in the cloud were not all the same weight. Lighter elements slowly formed into the huge outer planets, Jupiter, Saturn, Uranus and Neptune, and possibly Pluto as well. The inner planets, including Earth, were mostly formed from heavier elements, which drifted nearer to the centre of the cloud, and nearer to the sun.

Surface temperature: **-30°C to +40°C**
Radius: **6,378 km**

Rotation: **23 hours, 56 minutes, 4 seconds**

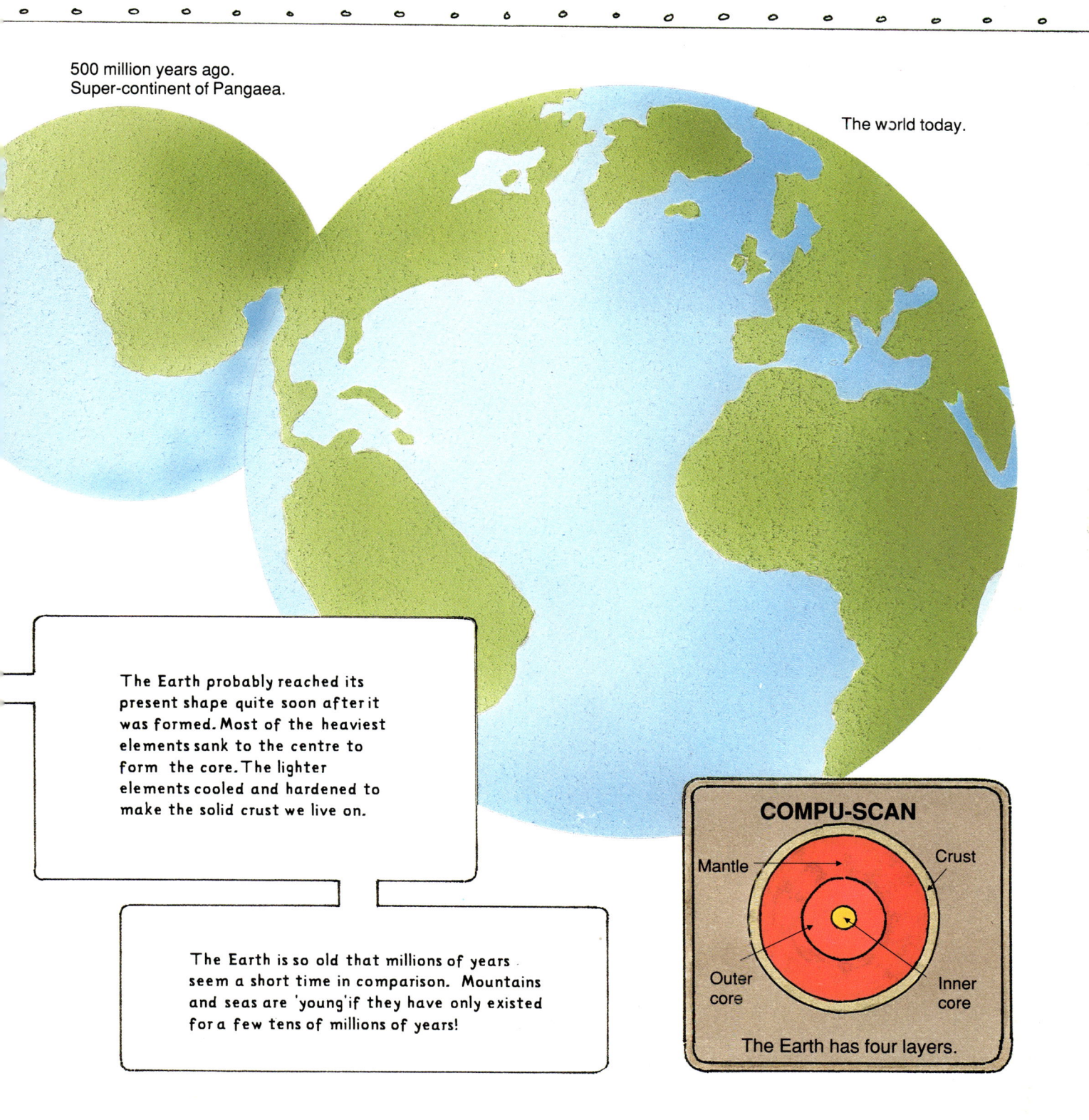

500 million years ago.
Super-continent of Pangaea.

The world today.

The Earth probably reached its present shape quite soon after it was formed. Most of the heaviest elements sank to the centre to form the core. The lighter elements cooled and hardened to make the solid crust we live on.

The Earth is so old that millions of years seem a short time in comparison. Mountains and seas are 'young' if they have only existed for a few tens of millions of years!

COMPU-SCAN

Mantle — Crust
Outer core — Inner core

The Earth has four layers.

The crust forms a thin skin over the surface of the Earth. Seventy percent of the crust is covered in water by the oceans. Under the oceans the crust can be as little as 4 km deep.

Four kilometres! I could walk that in 2 hours.

However, the crust under continents is much thicker. Continents are the large masses of dry land, like Europe or America, on which we live.

Under the continents the crust is an average of 35 to 40 km thick. Because the crust under the continents is heavier, it presses harder into the mantle beneath.

Ocean

Continent

Crust   minimum 4 km

35 to 40 km

Pressure

Mantle

COMPU-MAP

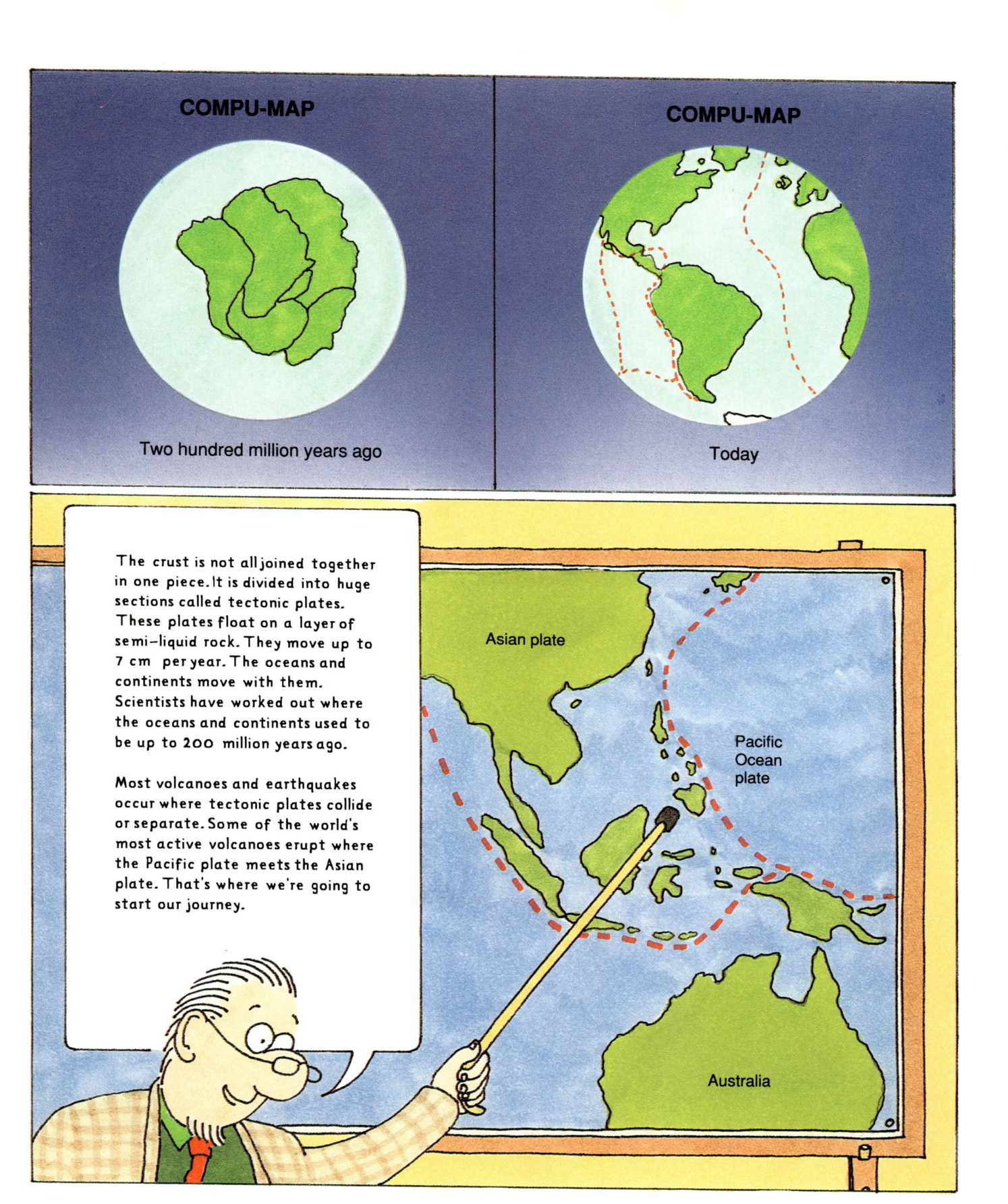

Two hundred million years ago

COMPU-MAP

Today

The crust is not all joined together in one piece. It is divided into huge sections called tectonic plates. These plates float on a layer of semi-liquid rock. They move up to 7 cm per year. The oceans and continents move with them. Scientists have worked out where the oceans and continents used to be up to 200 million years ago.

Most volcanoes and earthquakes occur where tectonic plates collide or separate. Some of the world's most active volcanoes erupt where the Pacific plate meets the Asian plate. That's where we're going to start our journey.

Asian plate

Pacific Ocean plate

Australia

Destination: **MOUNT APO, MINDANAO**

We're going to enter the Earth down this volcano.

Activate force field.

Temperature 500 degrees Centigrade and rising.

Oh no!

At high temperatures rocks melt and form a liquid called *magma*. In some places there are holes in the Earth's crust where magma escapes into the open air. These holes are called *volcanoes*.

COMPU-DATA

MAP

DATA

SCAN

Magma chamber

Volcanic ash often piles up to form a *cinder cone*. This is the familiar shape of volcanoes.

Longitude: **127° East**
Latitude: **6° North**

Height: **2,954 m**
Warning: *** **active volcano** ***

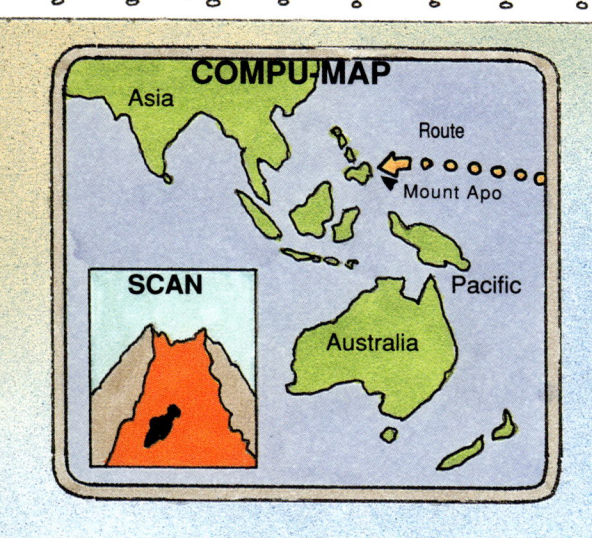

COMPU-MAP

Asia · Route · Mount Apo · Pacific · Australia

SCAN

Sometimes magma is very liquid and seeps out of a volcano slowly and smoothly. These slow volcanoes are found under the oceans and where tectonic plates are moving away from each other.

If magma is sticky, it gets bottled up inside a volcano. This stops the magma from escaping. Pressure builds up until it becomes so great that the sticky magma explodes. A huge cloud of dust, steam and liquid rock bursts outwards. These kind of volcanic eruptions are among the most violent events in the world.

COMPU-DATA

The word 'volcano' comes from 'Vulcan'. Vulcan was the Roman god of fire.

9

Destination: **JAPANESE ISLAND ARC** Western Pacific

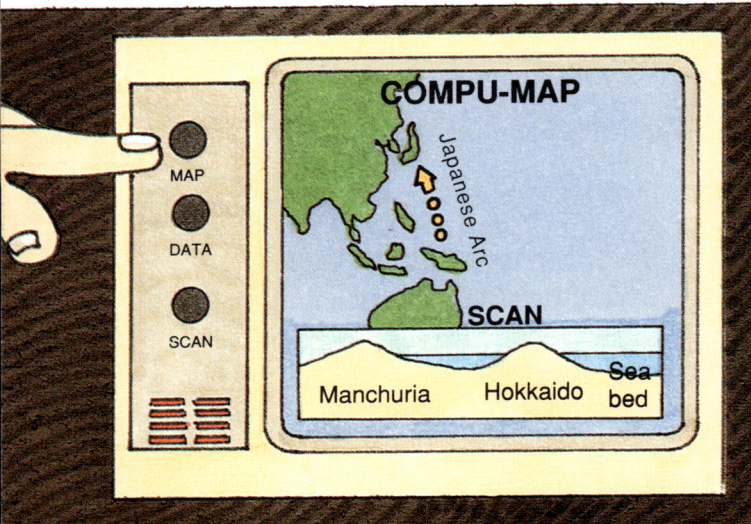

COMPU-MAP

- MAP
- DATA
- SCAN

Japanese Arc

SCAN

Manchuria — Hokkaido — Sea bed

COMPU-DATA
MERCALI 9 - TOTAL PANIC!

The *Richter Scale* measures the intensity of earthquake tremors. The *Mercali Scale* shows the effect they have on the surface.

Japan is a very active earthquake zone. The Japanese arc lies on the boundary between the Pacific and Eurasian tectonic plates. Like volcanoes, earthquakes are most common at the boundaries between the Earth's tectonic plates. The huge masses of rock which make up the plates are constantly trying either to move past or to move away from each other. Pressure builds up. Eventually, the plates move a bit and there is an earthquake. Areas where the crust is moving and shifting against itself in this way are called *fault lines*.

Longitude: **125° to 175° East**
Latitude: **30° to 50° North**

Earthquakes normally start at depths down to 300 km below the surface of the Earth. The centre of an earthquake is called its *focus*. The area on the surface directly above the focus is called the *epicentre*.

An earthquake tremor can be detected a long way from its focus - even on the other side of the world. Scientists can measure the speed at which the tremors travel through the Earth and the distance they travel. Over one million earthquakes happen every year. But most of them are so small that nobody feels them.

Everything's shaking!

My glasses!

We must be at the focus of an earthquake!

Let's get out of here!

That won't help!

11

Destination: # ROOTS OF THE HIMALAYAS

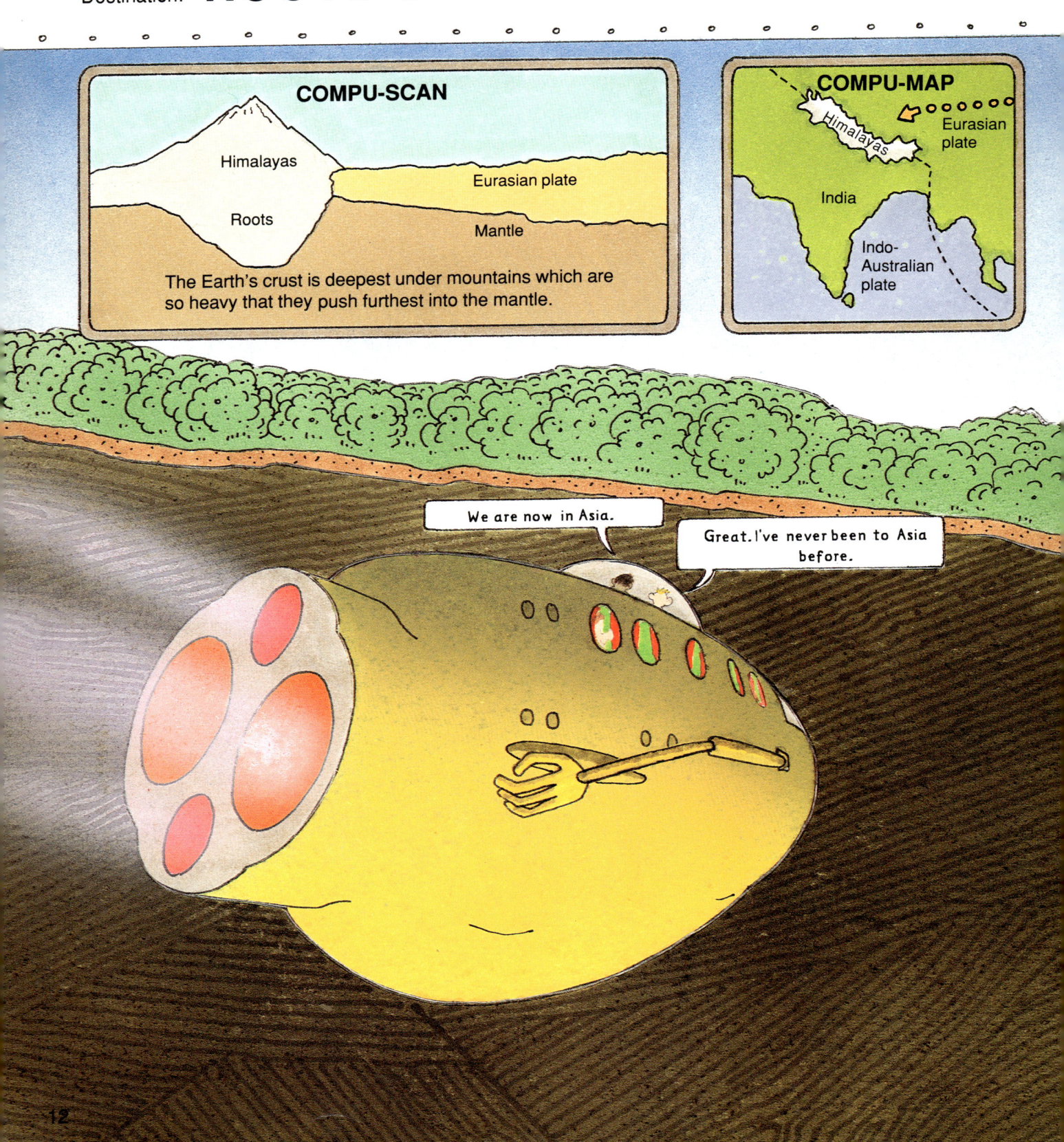

COMPU-SCAN

Himalayas

Roots

Eurasian plate

Mantle

The Earth's crust is deepest under mountains which are so heavy that they push furthest into the mantle.

COMPU-MAP

Himalayas
Eurasian plate
India
Indo-Australian plate

We are now in Asia.

Great. I've never been to Asia before.

Longitude: **75° to 90° East**
Latitude: **25° to 35° North**

Depth: **90 km**

Rock type: **Basalt**

Mountains are formed when tectonic plates collide together. The rocks at the edges of the plates slowly scrunch up against each other.

The Himalayas rise up where the Indian plate is pushing against the Eurasian plate. The Himalayas are the youngest mountains in the world. They are also the tallest. They have not yet been worn away by the weather or by further movements of the Earth's crust.

The crust is at its thickest underneath mountains. This is because the weight of a mountain pushes the lower crust deep into the mantle. The mass of the crust beneath a mountain is called its *root*.

The oldest mountain ranges, such as the Urals, are found at the centres of continents. Some old mountains are worn so flat that only the depth of their roots tells us how tall they once were.

Destination: # MID-ATLANTIC RIDGE

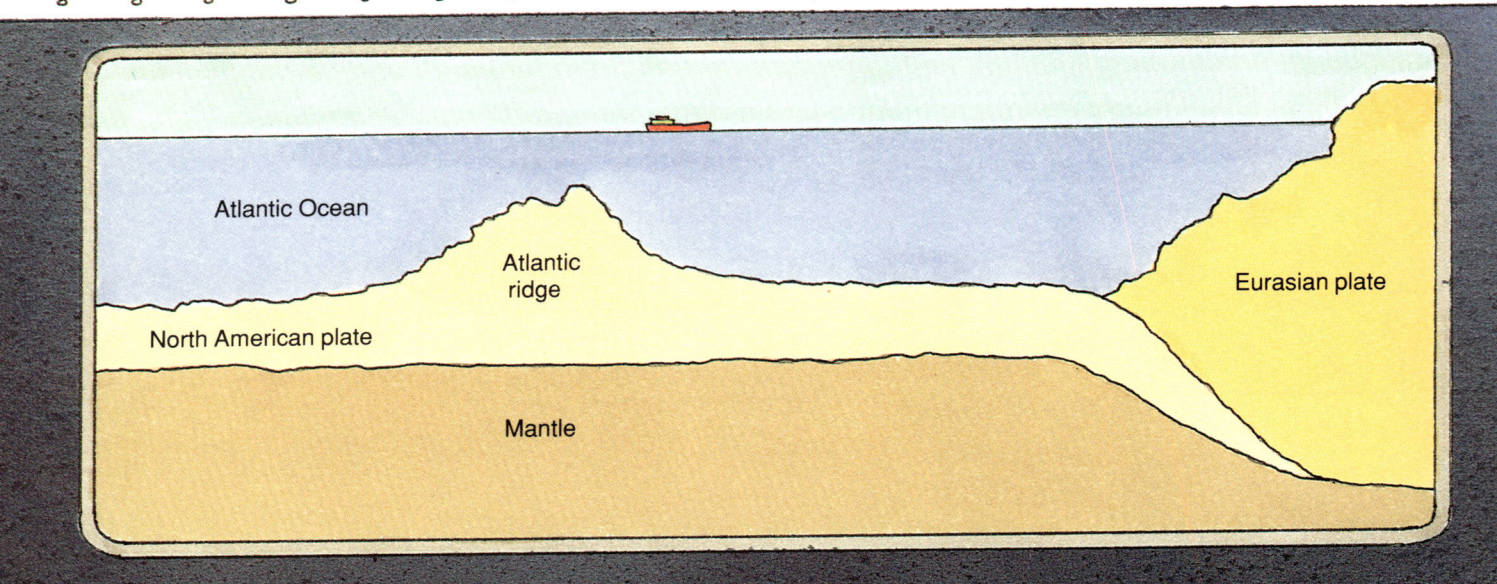

Diagram labels: Atlantic Ocean; Atlantic ridge; North American plate; Mantle; Eurasian plate

A range of underwater volcanoes runs down the centre of the Atlantic ocean. This underwater mountain range is called the Mid-Atlantic Ridge.

There are similar underwater volcanic mountains under every ocean. They are called *ocean ridges*. Ocean ridges cause most of the movement of the Earth's tectonic plates. The underwater volcanoes of the ocean ridges erupt slowly and release fresh magma (molten rock) onto the ocean floor. The new rock cools and hardens. It pushes the old ocean floor away on either side.

So we know that the Atlantic ocean first appeared as a crack in the Earth's surface about 190 million years ago. At the very edges of the oceans the ocean floor is pushed slowly down underneath the continental tectonic plates, where it turns back into magma.

The downward movement of the ocean floor at the edges pulls at the continental plates and causes them to move very slowly.

The deepest depths of the oceans are near the edges where the ocean floor starts to slope down beneath the continents. These depths are called *trenches*.

Longitude: **16° to 30° East**
Latitude: **58° South to 50° North**
Height: **7,800 m from greatest depth to highest peak**
Rock type: **Basalt**

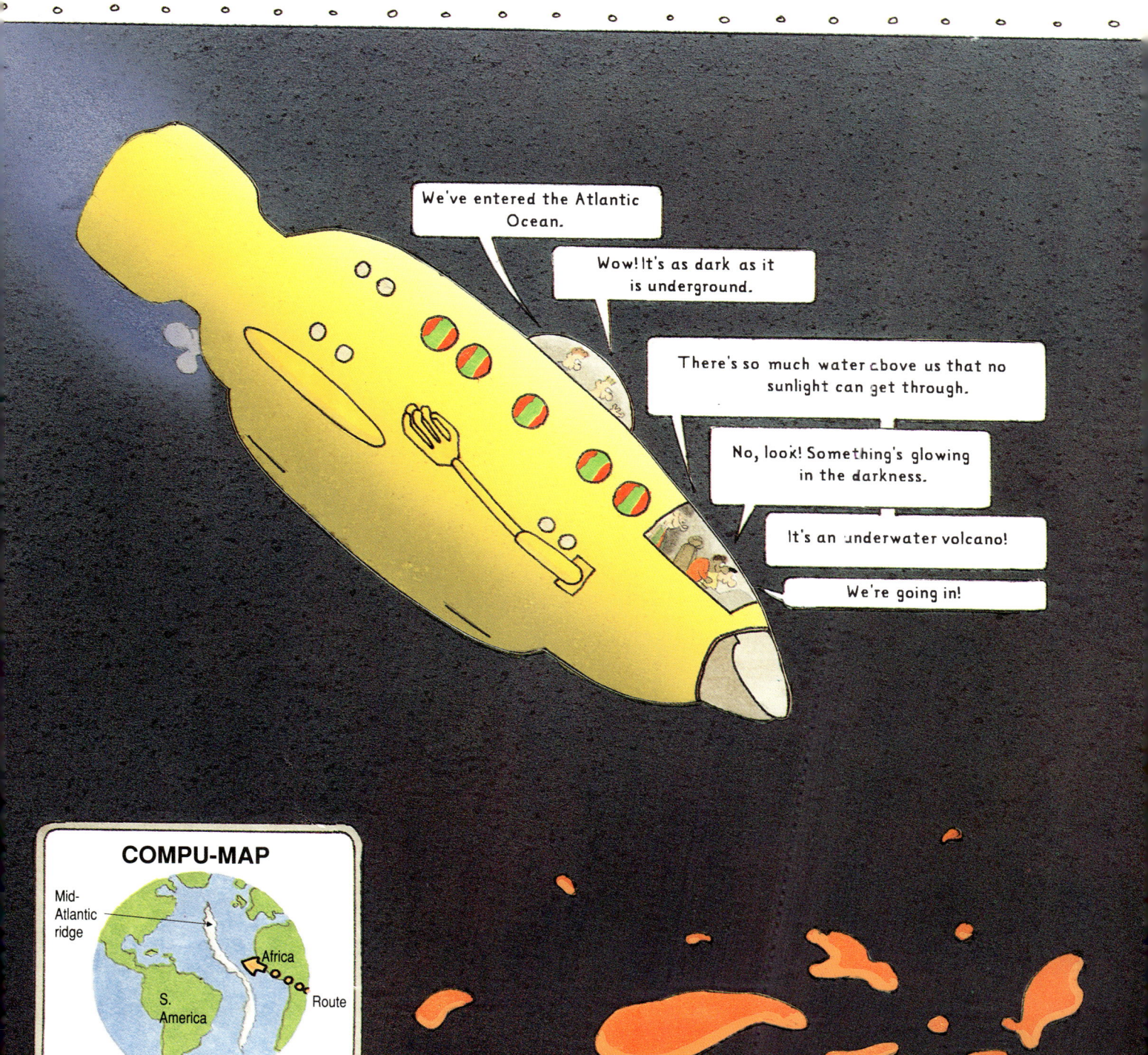

We've entered the Atlantic Ocean.

Wow! It's as dark as it is underground.

There's so much water above us that no sunlight can get through.

No, look! Something's glowing in the darkness.

It's an underwater volcano!

We're going in!

COMPU-MAP
Mid-Atlantic ridge
Africa
S. America
Route

Destination: # NORTH AMERICAN BASIN

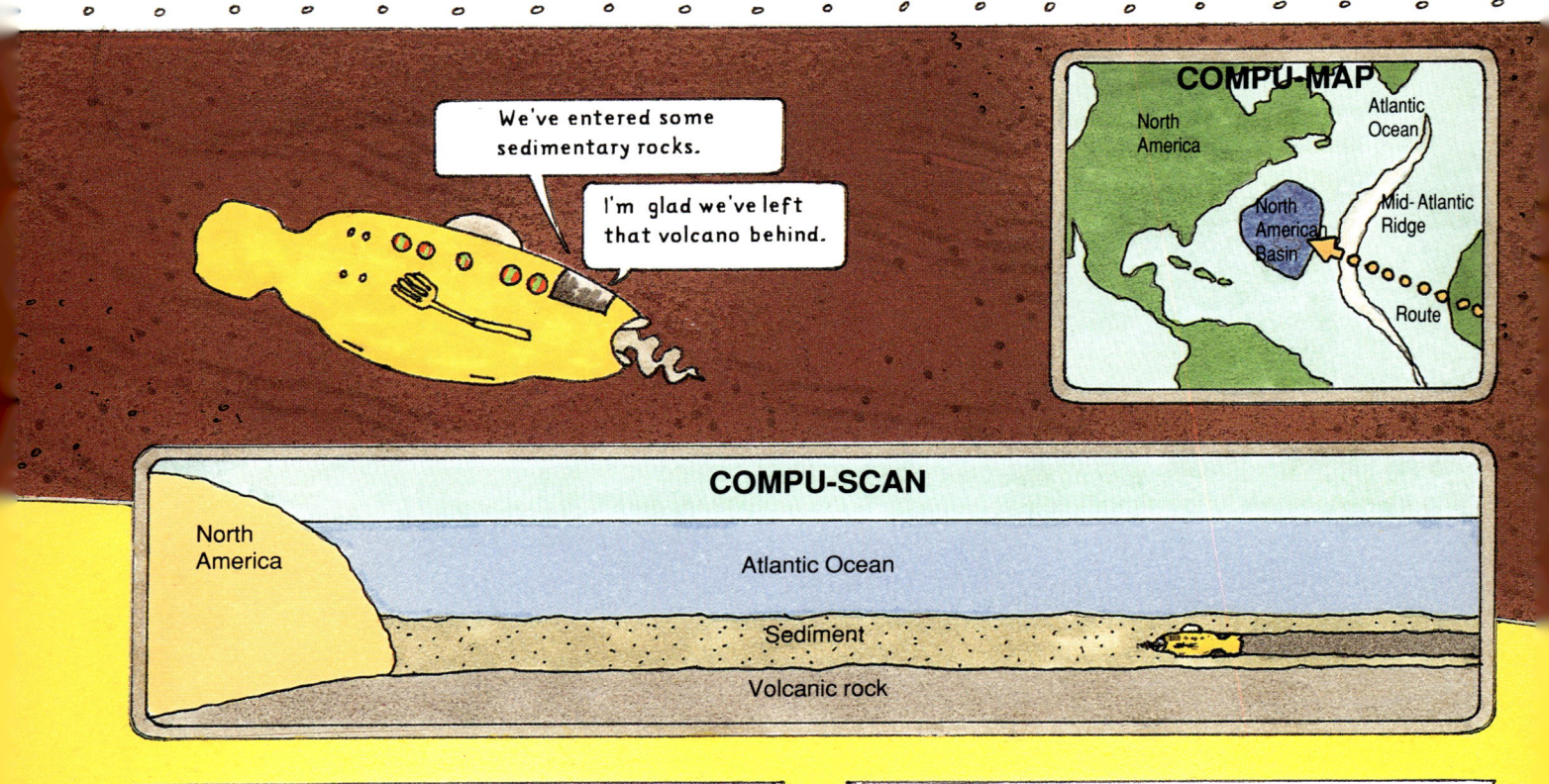

We've entered some sedimentary rocks.

I'm glad we've left that volcano behind.

COMPU-MAP
North America — Atlantic Ocean — North American Basin — Mid-Atlantic Ridge — Route

COMPU-SCAN
North America — Atlantic Ocean — Sediment — Volcanic rock

Beyond the Atlantic Ridge lies a flat area of ocean floor. These flat areas are called *Abyssal Plains*. They are flat because they are covered in *sediment*.

The rocks of the continents are constantly being worn away. They are worn away by heat, cold, ice, wind and rain. Small pieces break off. They are washed away by rain, or blown by the wind. Most of these grains of rock are carried down to the oceans by rivers.

This process of wearing away of rock is called *erosion*. The eroded rock ends up as sediment on the ocean floor. The sediments grow thicker and thicker. Over millions of years the sediment at the bottom is squeezed by the weight of sediment above. The squeezed, or compressed sediment turns into *sedimentary rock* such as *sandstone* or *shale*.

Longitude: **45° to 75° East**
Latitude: **20° to 40° North**

Average depth: **5,000 - 6,000 m**

COMPU-DATA

There are three main types of rock:

IGNEOUS - formed from magma.

SEDIMENTARY - formed from the compressed grains of other rocks or from the remains of tiny animals.

METAMORPHIC - formed by changing the other two by intense heat and pressure.

COMPU-DATA
Limestone caves

Limestone is a sedimentary rock. It is often formed from the shells of tiny creatures. Water washes it away making underground caves.

Dripping water leaves minerals which slowly grow to form stalactites (from above), and stalagmites (from below).

We are about to enter the United States of America.

We've speeded up!

Sedimentary rock can be quite soft. This machine cuts through it like butter.

Seat belts on!

Destination: COMO BLUFF, WYOMING, USA

When plants and animals die, they are sometimes preserved in sedimentary rock. The remains preserved in the rock are called *fossils*.

Scientists can tell the age of many rocks by the fossils they find within them. They know how long ago the plants and animals existed and so they can guess when the rocks were formed.
Fossils also show how animals and plants have changed over the years.

Only the hard parts of living things can be fossilised; for example, bones and shells. However, even soft-bodied animals can leave traces in the rocks. Their bodies may rot away and leave a hole where they used to be.

Evidence of the earliest life so far discovered is in the form of *cherts*. Cherts are traces left in crystal by tiny plants up to 3.3 billion years ago.

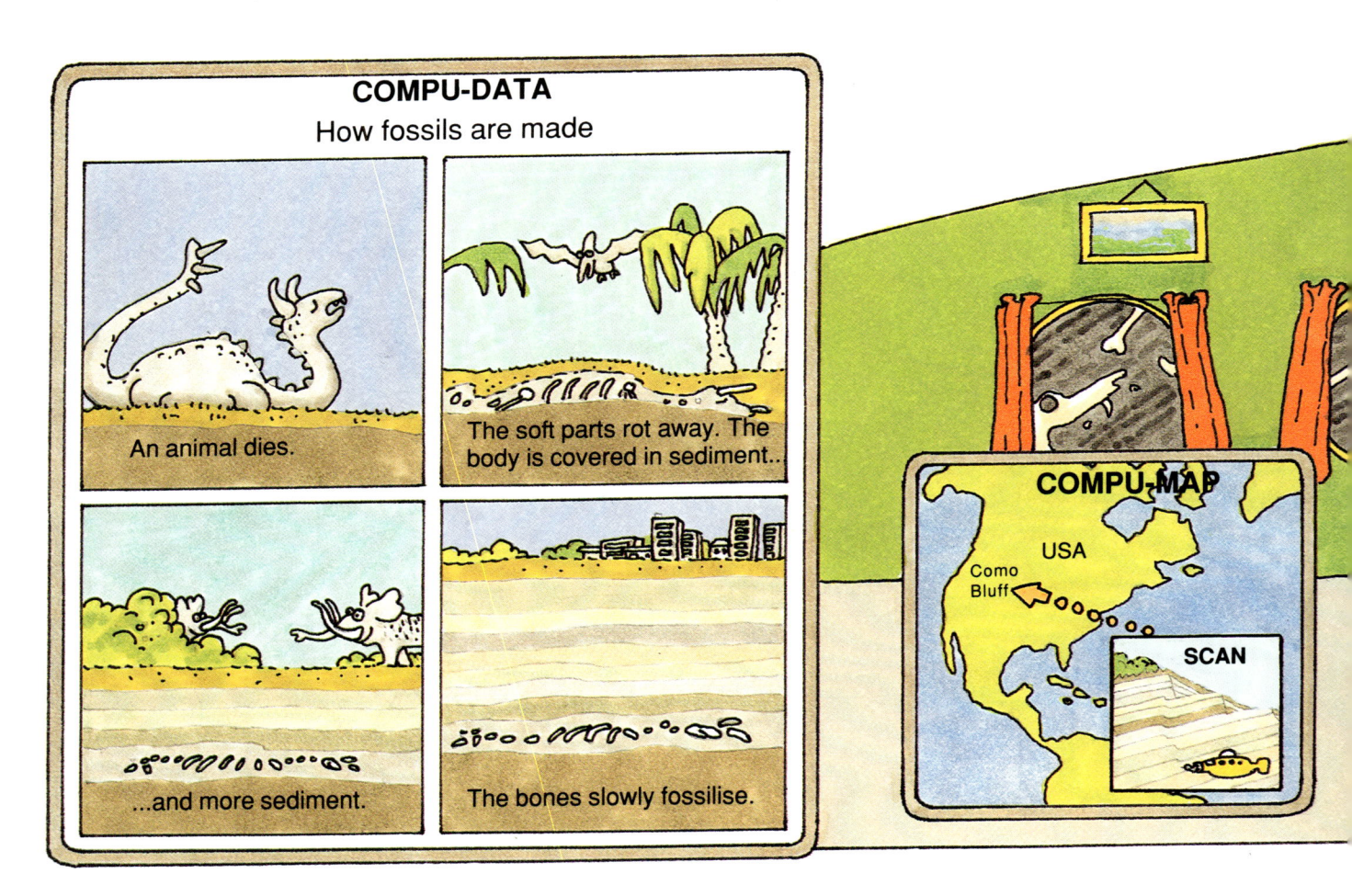

COMPU-DATA
How fossils are made

An animal dies.

The soft parts rot away. The body is covered in sediment..

...and more sediment.

The bones slowly fossilise.

COMPU-MAP

USA
Como Bluff

SCAN

Longitude: **104° West**
Latitude: **44° North**

Rock types: **Sedimentary – shale, sandstone, limestone**
Depth of formation: **400 m**

Dinosaurs flourished in the Jurassic period between 213 and 144 million years ago. Como Bluff, Wyoming, is one of the best places to find dinosaur bones.

At Como Bluff almost the entire history of the dinosaurs is preserved in layers of sedimentary rock.

COMPU-DATA

Every week between 1880 and 1890 more than a ton of dinosaur bones was excavated from Como Bluff.

It's the thigh bone of a Brontosaurus.

The biggest animal ever to walk the Earth.

Woof!

Destination: **ALBERTA BASIN, CANADA**

The Alberta Basin, beside the Rocky Mountains, is the major coal-producing area of Canada. Most coal was formed 300 million years ago when all the continents were joined together in the super-continent called *Pangaea*.

Between 360 and 268 million years ago, the seas around Pangaea rose. This period is called the *Carboniferous Period* after the Latin word *carbo*, meaning coal. Huge, marshy deltas formed between the land and the sea. Tropical forests grew on the marshy land around the equator. The remains of these ancient forests turned into coal.

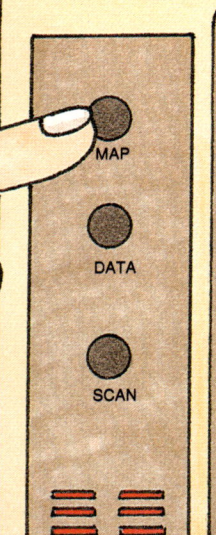

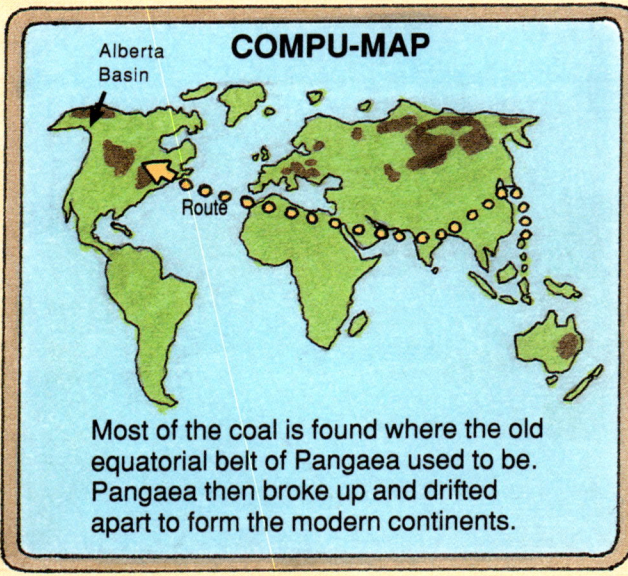

COMPU-DATA
How coal is formed

Dead leaves and wood turn into peat.

Muddy sediment
Lignite

Muddy sediment falls on the peat. The peat turns into *lignite* coal.

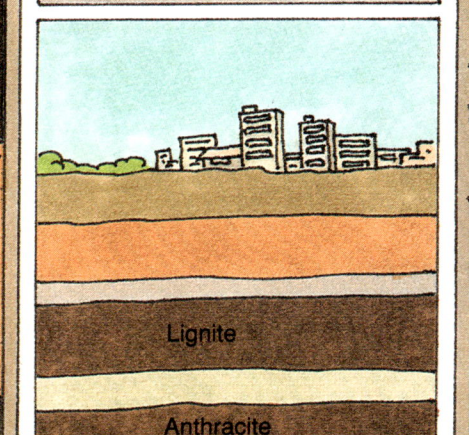

Lignite
Anthracite

More sediment falls and pressure increases. Lignite turns into *anthracite* coal.

COMPU-MAP

Alberta Basin
Route

MAP
DATA
SCAN

Most of the coal is found where the old equatorial belt of Pangaea used to be. Pangaea then broke up and drifted apart to form the modern continents.

Longitude: **120° West**
Latitude: **53° North**

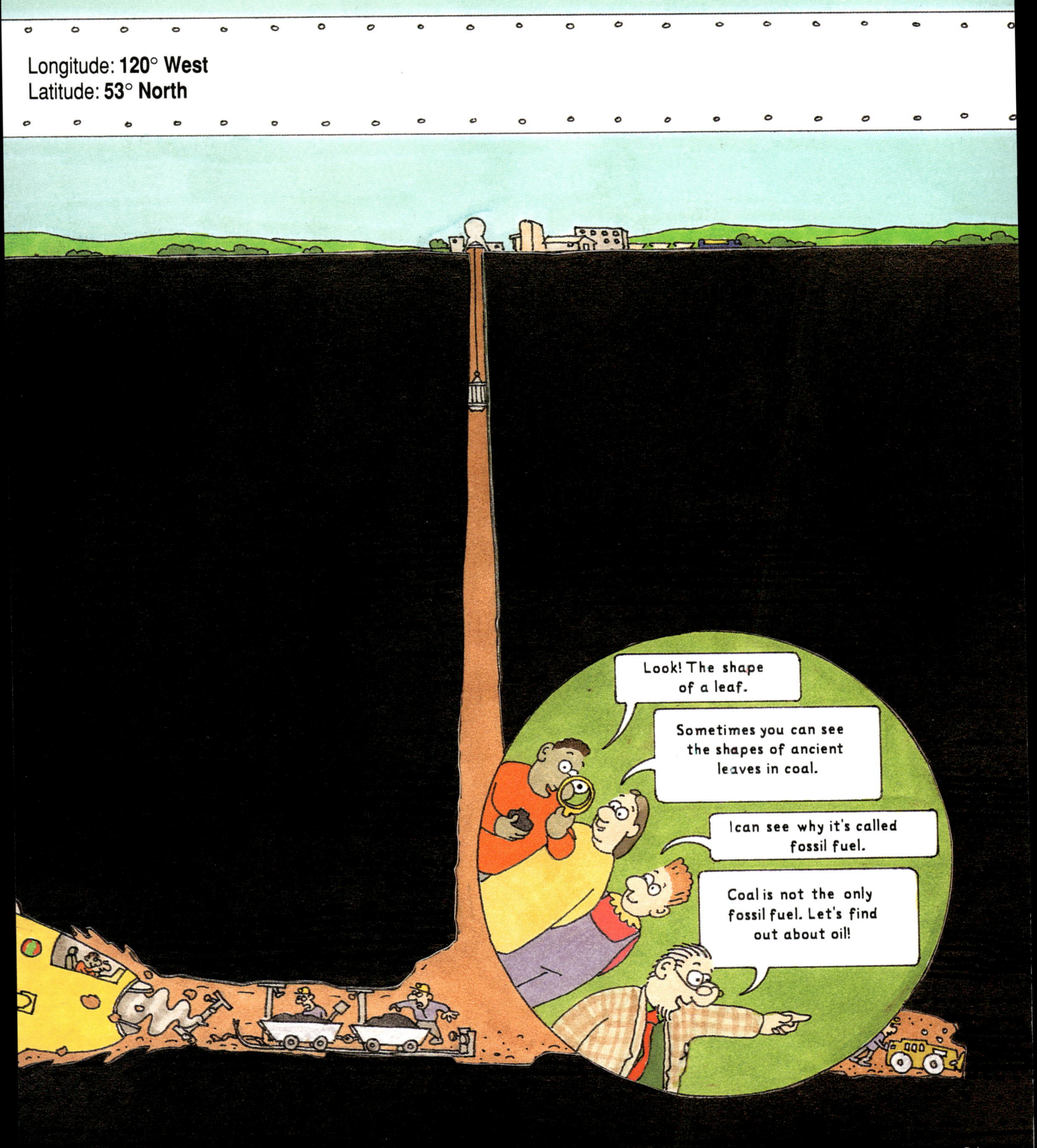

Destination: PRUDHOE BAY, ALASKA

There are large reserves of oil in Prudhoe Bay, Alaska. Oil is formed from the bodies of tiny plants called *phytoplankton*. Phytoplankton grow in shallow seas. Much of the oil we use today comes from phytoplankton which lived between 200 and 65 million years ago.

When phytoplankton die, they fall to the ocean floor. Sediment covers them.

After many years they rot. They turn into oil and gas. The thick, oozy oil soaks into the sedimentary rocks which have formed around it.

These rocks are often also soaked with water. Oil is lighter than water. So the oil slowly floats up through the water. It either reaches the surface, or, more usually, is trapped by harder rock. The oil gathers in this *trap*. From there it can be piped out. Oil is used to make petrol, plastic and other products.

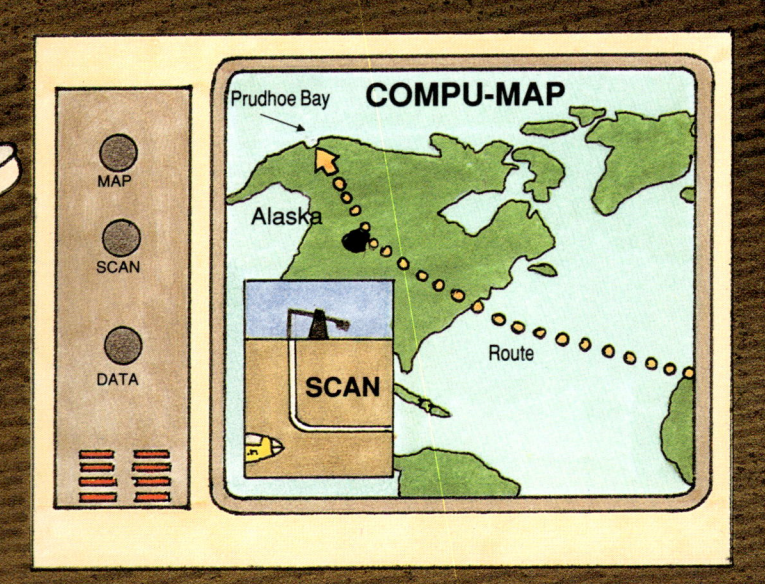

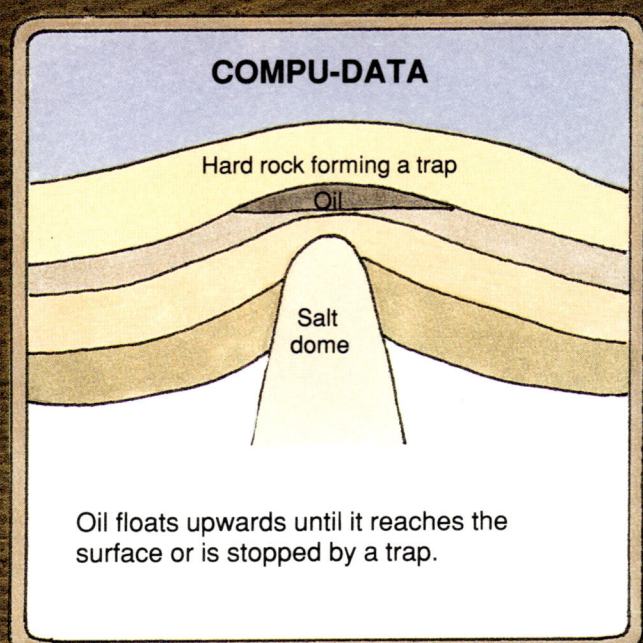

COMPU-MAP

COMPU-DATA

Oil floats upwards until it reaches the surface or is stopped by a trap.

Longitude: **149° West**
Latitude: **70° North**

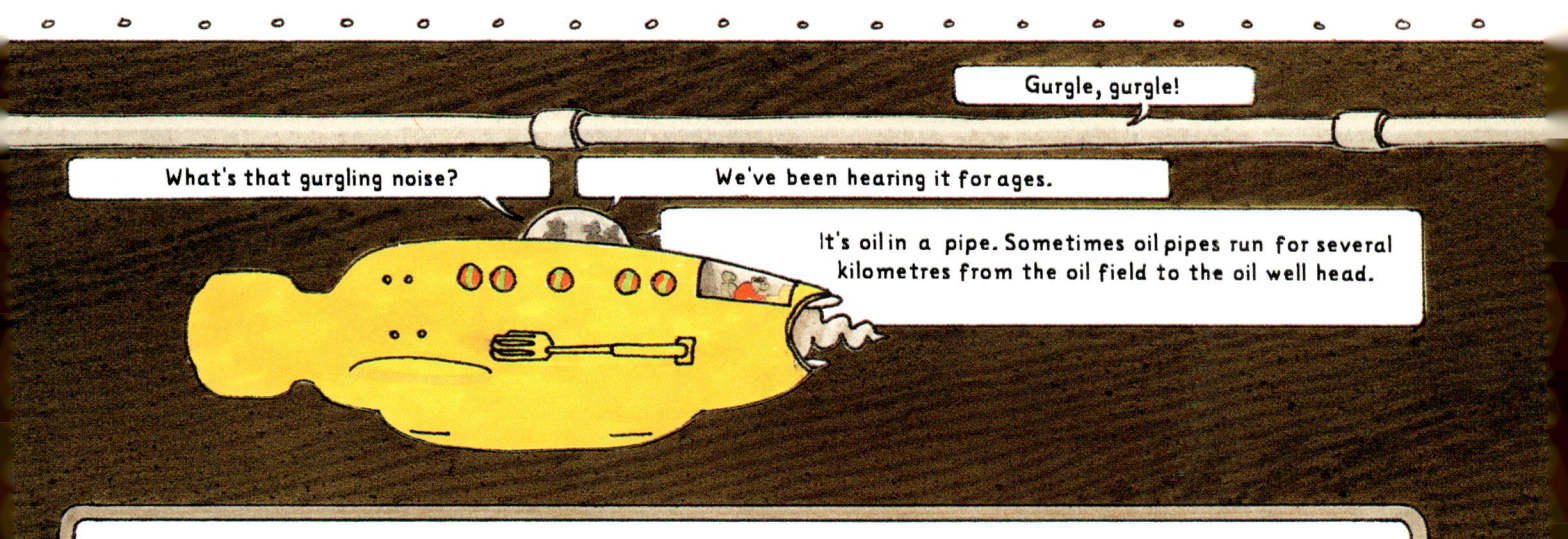

Gurgle, gurgle!

What's that gurgling noise?

We've been hearing it for ages.

It's oil in a pipe. Sometimes oil pipes run for several kilometres from the oil field to the oil well head.

COMPU-DATA
Variable gravity

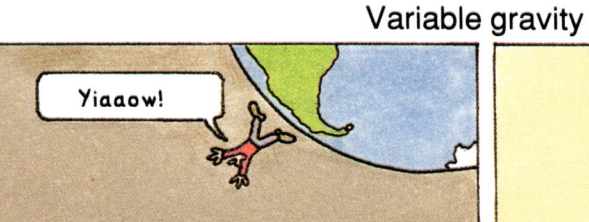

Yiaaow!

Gravity is a force which seems to pull things towards each other. Without gravity we would fall off the Earth.

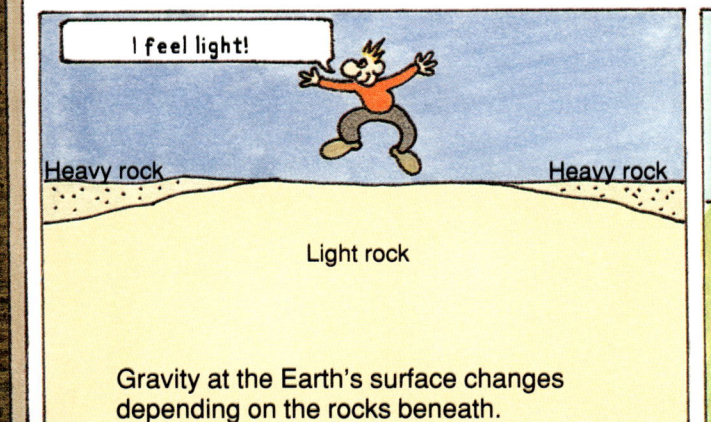

Heavy objects have stronger gravity than light objects. Some types of rock are heavier than others.

I feel light!

Heavy rock   Heavy rock

Light rock

Gravity at the Earth's surface changes depending on the rocks beneath.

Oil often collects above salt domes. Salt is light, so the gravity above it is low. Scientists hunt for oil by looking for areas of low gravity with a *gravimeter*.

Destination: CARLETONVILLE, SOUTH AFRICA

Metals are some of the most useful materials to be mined from the Earth's crust. They are used in a vast number of products, from huge ships to tiny computer parts. There are many metals. Iron, silver and gold are among the best known.

Like oil and gas, some metals are found in sedimentary rocks. Aluminium is one of these. But most metals are mined from igneous rock.

When igneous rock is forming from pools of liquid magma, metals floating within the magma sink to the bottom. There they collect in *deposits*.

Other metals seep into the surrounding rocks, where they solidify in layers called *veins*.

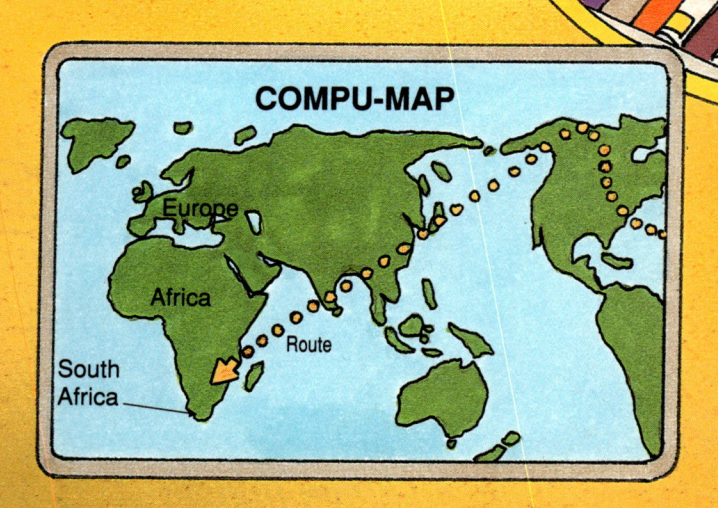

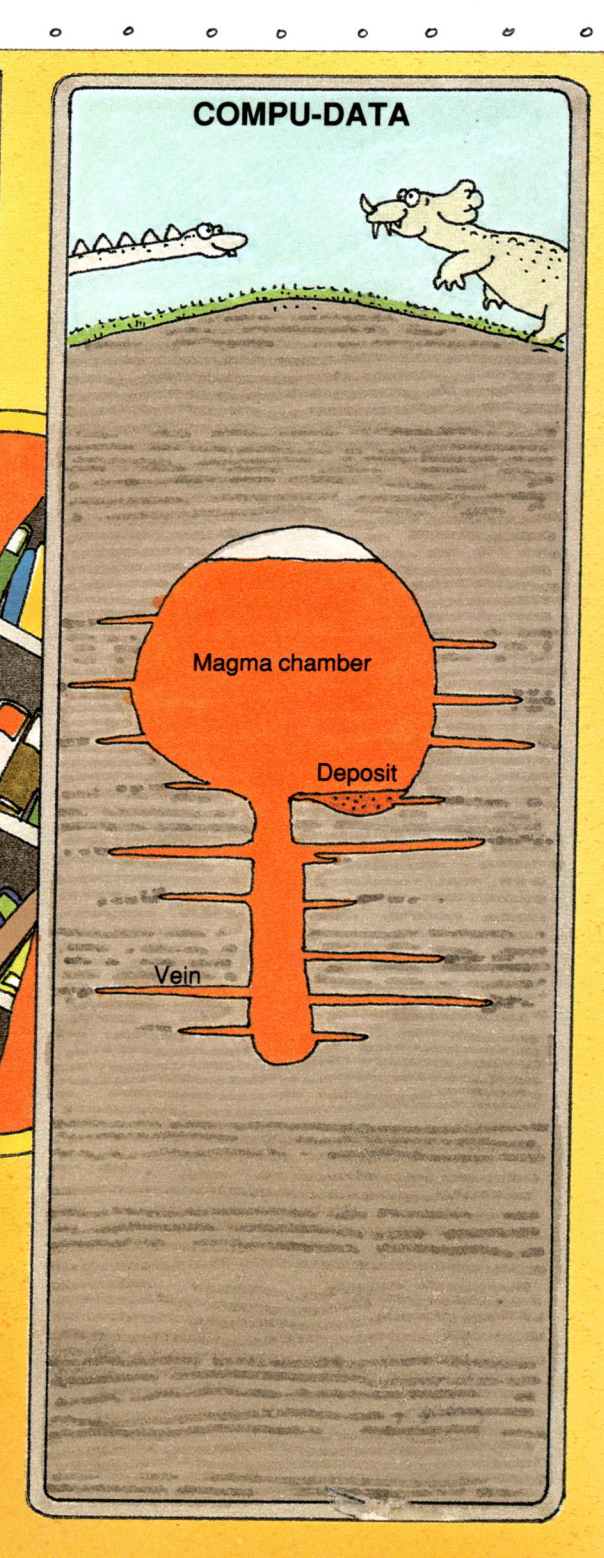

COMPU-DATA

Magma chamber
Deposit
Vein

COMPU-MAP

Europe
Africa
Route
South Africa

Longitude: **22° West**
Latitude: **27° South**

Depth: **3,777 m (2.34 miles)**
the deepest mine in the world

Gold is a very heavy metal. Magmas which have risen from very deep in the Earth's crust sometimes contain gold.

Ore from this type of magma is mined from the deepest mine in the world at Carletonville, South Africa.

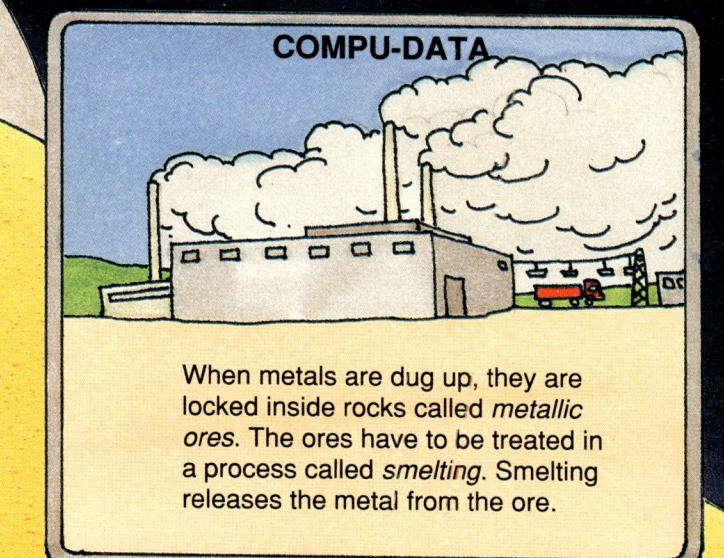

COMPU-DATA

When metals are dug up, they are locked inside rocks called *metallic ores*. The ores have to be treated in a process called *smelting*. Smelting releases the metal from the ore.

Is it gold?

I think so. But iron pyrites looks just the same. That's why it's called fool's gold.

Don't be foolish! It must be gold. This is a gold mine.

Some minerals come from even greater depths. Diamonds are often found in rocks which have risen from the mantle.

25

Destination: # THE MANTLE

Depth: Starts from between 4 and 70 km down to 2,900 km

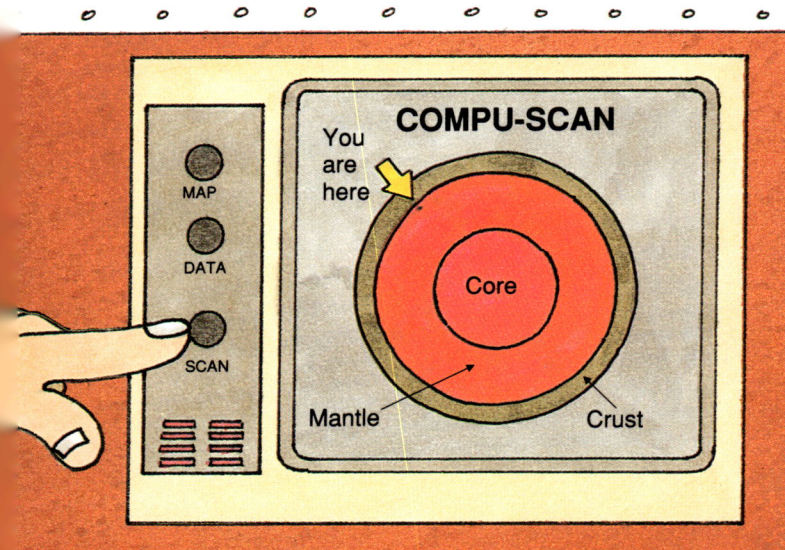

COMPU-SCAN

You are here →

Core · Mantle · Crust

Sixty percent of the Earth's mass is made up of the rocks of the mantle. Scientists can only estimate what kind of rock it is. The mantle is too deep to be drilled into.

The Earth's crust floats on top of the mantle. The boundary between crust and mantle lies at between 4 km under the oceans and up to 70 km under some mountains.

COMPU-DATA

Atmospheric pressure

All the air round the world is called the *atmosphere*. Even air has weight. At sea level, it weighs down on the Earth at a pressure of about 1 kg on every sq cm. Rock is much heavier than air. The pressure of rock within the Earth can be measured in millions of atmospheres.

Temperature: 1,100°C - 1,900°C
Pressure: Up to 1 million atmospheres

Compostion: Probably peridotite
State: Mainly solid

Just below the crust, the rocks of the mantle are almost molten. The tectonic plates which carry the continents and oceans slide around very slowly on top of this layer of semi-liquid rock.

Beneath the molten rocks the mantle becomes solid. This is at a depth of between 75 and 250 km.

Pressure increases towards the centre of the Earth. The rock is squeezed by the pressure. It becomes denser and heavier.

COMPU-DATA

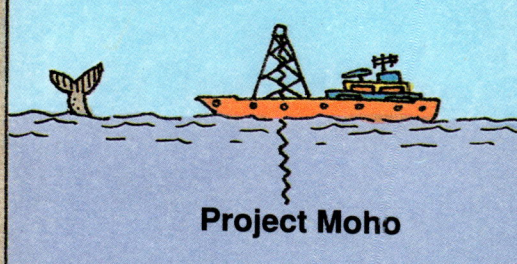

Project Moho

In 1957 a plan was made to drill down to the mantle. A test drill was made beneath 3 km of ocean off the coast of California. The project was later abandoned.

Pressure one million atmospheres. Activate booster force field!

Temperature 3,000 degrees Centigrade and rising. Activate the refrigeration unit.

This is nothing. Wait until we get to the core!

Destination: **THE CORE**

Depth: **3,420 km from the centre to the boundary with the mantle**

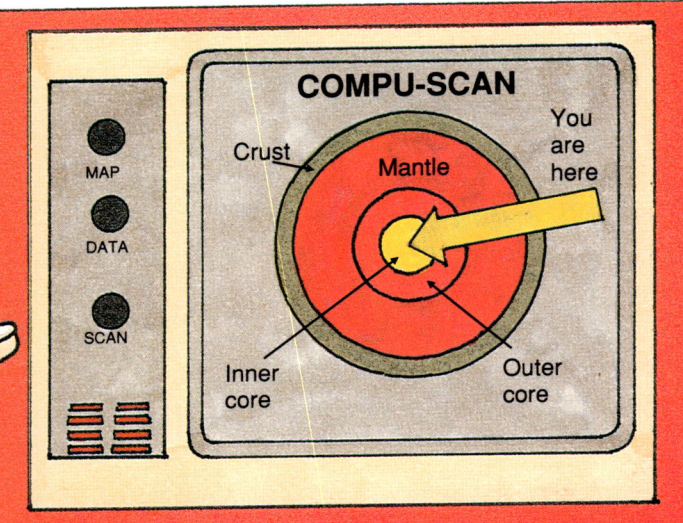

COMPU-SCAN

Crust • Mantle • You are here • Inner core • Outer core

MAP | DATA | SCAN

The Earth's core is divided into an inner core and an outer core. It is very hot. The temperature at the boundary between the outer core and the mantle is 1,900°C and at the centre of the Earth, the temperature rises to 4,500°C.

The pressure is also very great, rising from one million atmospheres in the outer core to 36 million atmospheres at the centre.

The Earth's outer core is liquid. The inner core is solid despite its very high temperature. This is because the very high pressure at the centre stops the inner core from melting.

The core is mainly made up of iron, nickel and oxygen. In addition, there is between 5% and 20% of some light materials. Scientists think that this may be silicon, sulphur or carbon.

Under very high pressures carbon forms into diamond. So perhaps some of the Earth's core is made of diamonds!

Temperature: **1,900°C - 4,500°C**
Pressure: **Up to 36 million atmospheres**

Composition: **48% oxygen, 37% metals, 12% silicon, sulphur or carbon**

I'm hot!

Activate scoop! Let's search for diamonds.

I don't know. There's no gravity at the centre of the Earth.

Which way is up?

Destination: **HOME!**

Distance: **6,378 km from the centre of the Earth**

The Earth was formed about 4.5 billion years ago. Heat from its formation is still held in the core. The heat can only escape very slowly through the huge mass of the mantle.

But eventually the heat will be gone, and the Earth will be as cold as the Moon. After that, the Sun will expand and swallow up the Earth.

But none of this will happen for many millions of years. In the meantime Professor Verne and the children have returned to the Earth's surface...

We made it!

I'm freezing!

Quick! Turn off the refrigeration unit.

Earth is much cooler at the surface.

Most of the heat there comes from the sun, not the Earth. Thank goodness it's a sunny day.

Right kids! Now that you've found out about the Earth, let's see how much you can remember.

Professor Verne's 'Centre of the Earth' quiz

1. *What is magma?*
   a) a gun
   b) a female magpie
   c) liquid rock

2. *What is the mantle?*
   a) an overcoat
   b) the mass of Earth between crust and core
   c) a ledge above a fireplace

3. *What is Pangaea?*
   a) a super-continent
   b) a girl's name
   c) the muck in your frying pan

4. *What is sediment?*
   a) a type of cement
   b) a layer of pieces of eroded rock
   c) an emotion

5. *What are phytoplankton?*
   a) microscopic plants
   b) very small surf boarders
   c) a spelling mistake

6. *What is the crust?*
   a) the hard bit of your sandwich
   b) the dirt under your fingernails
   c) a layer of solid rock at the surface of the Earth

Answers: 1) c. 2) b. 3) a. 4) b. 5) a. 6) c (and a, of course).

INDEX

Abyssal plain   16
Alberta basin   20
Anthracite   20
Asia   12
Asian plate   7
Atlantic ocean   15
Atmosphere   26
Basalt   13, 15
Brontosaurus   19
Carboniferous   20
Carletonville   24, 25
Cherts   18
Cinder cone   8
Coal   20
Como bluff   18
Continent   6
Crust   5, 6, 25, 26
Diamond   25, 28, 29
Dinosaur   19
Earthquakes   7, 10, 11
Elements   4, 5
Epicentre   11
Equator   20
Erosion   16
Fault line   10
Focus   11

Fools gold   25
Fossil fuel   21
Fossils   18
Gold   25
Gravity   23, 29
Gravimeter   23
Himalayas   13
Igneus   17, 24
Inner core   5, 28, 30
Jurassic   19
Lignite   20
Limestone   17, 19
Magma   8, 9, 14, 24, 25
Mantle   5, 6, 25, 26
Mercali scale   10
Metals   24, 25
Metamorphic   17
Mid Atlanic ridge   14, 16
Moho   27
Mount Apo   8
North American Basin   16
Oceans   6, 7, 14
Ocean floor   14, 16
Ocean ridge   14
Oil   22
Ore   25

Outer core   5, 28, 30
Pacific plate   7
Pangea   5, 20
Peat   20
Petrol   22
Phytoplankton   22
Planet   4
Plastic   22
Richter scale   10
Rotation   5
Root   12, 13
Salt dome   23
Sandstone   16, 19
Sediment   16, 17, 24
Shale   16
Shells   18
Smelting   25
Stalactites   17
Stalagmites   17
Tectonic plates   7, 9-10, 13-24, 27
Trap   22
Trenches   14
Urals   13
Volcanoes   7 – 9, 14
Veins   24, 28

British Library Cataloguing in Publication Data
Johnson, Kipchak
  Centre of the Earth. - (Fantastic Journey Series)
  I. Title II. Series
  551.1
ISBN 0 340 57082 2
First published 1992
© 1992 Kipchak Johnson/Lazy Summer Books

The rights of Kipchak Johnson to be identified as the author of the text of this work has been asserted by him in accordance with the Copyright, Designs and Patents Act 1988.

All rights reserved. No part of this publication may be reproduced or transmitted in any form or by any means, electronic or mechanical, including photocopy, recording, or any information storage and retrieval system, without permission in writing from the publisher or under licence from the Copyright Licensing Agency Limited. Further details of such licences (for reprographic reproduction) may be obtained from the Copyright Licensing Agency Limited, of 90 Tottenham Court Road, London W1P 9HE.

Printed in Great Britain for the educational publishing division of Hodder and Stoughton Ltd, Mill Road, Dunton Green, Sevenoaks, Kent by Cambus Litho Limited, East Kilbride.